AF270199

Gavials

by Grace Hansen

Abdo Kids Jumbo is an Imprint of Abdo Kids
abdobooks.com

abdobooks.com

Published by Abdo Kids, a division of ABDO, P.O. Box 398166, Minneapolis, Minnesota 55439.
Copyright © 2021 by Abdo Consulting Group, Inc. International copyrights reserved in all countries.
No part of this book may be reproduced in any form without written permission from the publisher.
Abdo Kids Jumbo™ is a trademark and logo of Abdo Kids.

Printed in the United States of America, North Mankato, Minnesota.

052020

092020

THIS BOOK CONTAINS
RECYCLED MATERIALS

Photo Credits: Alamy, AP Images, iStock, Minden Pictures, Shutterstock

Production Contributors: Teddy Borth, Jennie Forsberg, Grace Hansen
Design Contributors: Dorothy Toth, Pakou Moua

Library of Congress Control Number: 2019956541
Publisher's Cataloging-in-Publication Data

Names: Hansen, Grace, author.
Title: Gavials / by Grace Hansen
Description: Minneapolis, Minnesota : Abdo Kids, 2021 | Series: Spooky Animals | Includes online
    resources and index.
Identifiers: ISBN 9781098202507 (lib. bdg.) | ISBN 9781098203481 (ebook) | ISBN 9781098203979
    (Read-to-Me ebook)
Subjects: LCSH: Crocodiles--Juvenile literature. | Crocodiles--Behavior--Juvenile literature. | Reptiles—
    Behavior--Juvenile literature. | Curiosities and wonders--Juvenile literature.
Classification: DDC 596.018--dc23

Table of Contents

Gavials

Gavials are members of the **crocodilian** family. They can be found in northern India and Nepal. However, their range was once much greater.

5

Gavials live in freshwater rivers.

Because of their weak legs,

they rarely leave the water.

When they do, it is to warm

themselves or lay eggs.

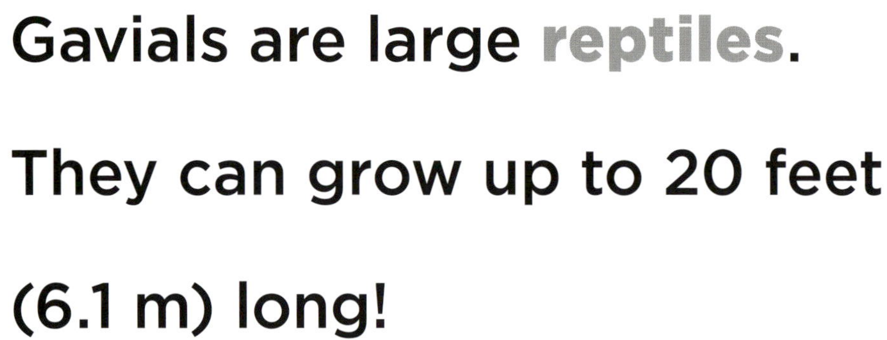

Gavials are large **reptiles**.

They can grow up to 20 feet

(6.1 m) long!

They often weigh around 350 pounds (160 kg). But some have weighed up to 2,000 pounds (907 kg)!

Gavials look much like other

crocodilians. But their long,

skinny **snout** sets them apart.

gavial

saltwater crocodile

black caiman

Nile crocodile

Cuvier's dwarf caiman

American crocodile

spectacled caiman

American alligator

While it might not look it, a gavial's jaws are deadly! Its mouth holds more than 100 extremely sharp teeth.

Hunting & Food

The gavial sweeps its mouth

sideways to catch its **prey**.

Adults mainly eat large fish.

Young gavials eat insects,

frogs, and other small animals.

17

Baby Gavials

Gavials spend most of their time alone. However, they will gather to nest in the spring. Females lay around 30 to 50 eggs. About 3 months later, the eggs hatch.

Young gavials stay near their
mothers. After a few months,
they are big and strong enough
to be on their own.

More Facts

- The exact lifespan of the gavial is not known. It is thought that they live for around 50 to 60 years in the wild. This is the same as other crocodilian species.

- Gavials have existed on Earth for more than 200 million years. They have changed very little in that time.

- Gavials are also known as Gharials. They are critically endangered.

Glossary

critically endangered – in extremely high danger of becoming extinct.

crocodilian – a reptile of the group that includes the crocodile. Alligators, caimans, and gavials are also crocodilians.

prey – an animal that is hunted and eaten by another animal for food.

reptile – a cold-blooded animal with a skeleton inside its body and dry scales or hard plates on its skin. Most reptiles lay eggs.

snout – the projecting nose and mouth of an animal.

Index

Abdo Kids ONLINE
FREE! ONLINE MULTIMEDIA RESOURCES

Visit **abdokids.com** to access crafts, games, videos, and more!

Use Abdo Kids code **SGK2507** or scan this QR code!